YOUR KNOWLEDGE HAS

Congestion Charges. International Experiences and Suggestions for Germany

Bibliographic information published by the German National Library:

The German National Library lists this publication in the National Bibliography; detailed bibliographic data are available on the Internet at http://dnb.dnb.de.

ISBN: 9783346441676
This book is also available as an ebook.

© GRIN Publishing GmbH
Nymphenburger Straße 86
80636 München

Print and binding: Books on Demand GmbH, Norderstedt, Germany
Printed on acid-free paper from responsible sources.

The present work has been carefully prepared. Nevertheless, authors and publishers do not incur liability for the correctness of information, notes, links and advice as well as any printing errors.

GRIN web shop: https://www.grin.com/document/1034382

Congestion Charges: International Experiences and Recommendations for Germany

Term Paper

Zeppelin University

242141 Matching Strategies for Sustainable Mobility

Friedrichshafen, 15.10.2019

CONTENTS

LIST OF FIGURES

LIST OF TABLES

Table of abbreviations

α	Costs of time (slope of elasticity)
τ	Congestion charge
ALS	Area Licensing Scheme
LEZ	LEZ
SGD	Singapore dollar
TfL	Transport for London
ULZ	Ultra Low Emission Zone

1. INTRODUCTION

"The future of mankind lies in the cities."

— Kofi Annan, Secretary-General United Nations

As part of the Paris Climate Agreement, 197 countries agreed to limit global warming to below 2 degrees Celsius compared to the pre-industrial level. This requires substantial measures to reduce carbon dioxide emissions. However, these plans are offset by a growing global gross domestic product, increasing urbanisation and, as a consequence, a high demand for fossil fuels. One direct result of these developments is congestion. Worldwide, congestion is a major problem in cities: a constant urbanisation coupled with more cars per inhabitant results in inefficiencies that materialise as congestion. In 2018, Berlin residents spent an average of 154 hours in traffic jams (Cookson, 2018). The personal annoyance is accompanied by problems such as increased environmental pollution, reduced productivity and other external effects. This means an average congestion rate of 14% and results in a total cost of €6.90 bn. for the city and a total cost of €2,811 for the individual driver. Especially in large cities there are various substitution possibilities in urban areas: The congestion problem could therefore be alleviated by switching to more space-efficient means of transport such as local public transport. The majority of cities lack incentive systems that encourage individual users to switch. This paper addresses this problem and discusses the possibility of introducing a price mechanism within the framework of a congestion charge. In the second chapter, transport economic models and the urban modal split are used as the basis. In the second half of the second chapter, congestion is considered as an expression of a collective action problem and possible counter-mechanisms. With regard to the research question, attention is paid to the congestion charge. In the third chapter, the case study method is used to analyse the three cities Stockholm, London and Singapore that have already introduced congestion charges. From the different price models as well as effects on the demand for transport services, recommendations for action for the application in German cities are developed in the fourth chapter. The last chapter contains a critical discussion and a conclusion.

2. THE TRANSPORT MARKET IN URBAN AREAS

2.1 Mobility in urban areas

The public transport demand is a derived demand resulting from the satisfaction of needs (Holmgren, 2007). Urban mobility is associated with the advantages of individual freedom, social inclusion as well as satisfaction of a society (Klinger et al., 2013). Individual

1

mobility is crucial when it comes to access to the labour market (Pilegaard & Fosgerau, 2008), education (Hine & Mitchell, 2017), health care and food (Clifton, 2004). Another dimension of mobility is an increased well-being which is differentiated in autonomy (self-determination as well as the ability to retain individuality and personal standards), personal growth and positive relations with others (Ryff, 1989). Stanley et al. (2011) emphasise that mobility fosters these categories. The benefits of mobility patterns are contrasted by decisionn parameters of the individual. According to Redman et al. (2013) the attributes can be differentiated into *physical* (for example reliability, frequency, speed, accessibility, price, information provision, ease of transfers / interchanges, vehicle condition) and *perceived* (comfort, safety, convenience, aesthetics).[1]

Direct factors on urban transport patterns may be the price of the ticket or the cost of fuel for the vehicle. Indirect factors are travel time, which can be expressed in opportunity costs, and punctuality, which can be expressed in delay costs (Lindsey & Verhoef, 2001). Vickrey (1969) argues that each individual has a preferred time of arrival t_i^* where *schedule delay costs* are present when the individual does not manage to arrive in time. These costs costs are denoted as $D_i(t - t_i^*)$ (Lindsey & Verhoef, 2000; Vickrey, 1969). According to Lindsey & Verhoef (2000), the costs for the individual (i) arriving at a certain time (t) can be expressed with the cost function $C_i(t) = \alpha_i T(t) + D_i(t - t_i^*) + c_i$ where $T(t)$ is the final trip duration and α_i is the value of time for each individual which can be understood as costs of opportunity. c_i are the direct costs for the individual.

Urban mobility patterns are framed by the infrastructure which is the supply for mobility (Vickrey, 1969). Especially in urban areas, the supply in the example of rail-services or the roads which are contested by bus services, motorised private transport and bicycles, is relatively inelastic because of high investment, a lack of available space and public acceptance (Flyvbjerg et al., 2004). Rietveld (1994) argues that the transport infrastructure is the basis for generalised transport costs which have an impact on the individual choice of the means of transport (movement of passengers), the accessibility of locations as well as the productivity of households and firms. Consequently, the available infrastructure determines the capacity and position of households decisions. These decisions are manifested in the modal split expressing the transport volume on different means of transport. If one assumes perfect rationality of the travellers, the modal split reflects the minimisation of the individual costs of travel. Donald et al. (2014) emphasise that the choice of means of transport is conditionally financial. Non-financial aspects such as social, moral and descriptive norms as well as ecological considerations are incorporated into the traveller's behaviour. Klöckner & Matthies (2004) argue that individual habits have a strong influence on individual transportation patterns. Accordingly, the decision-making process is subject

---

[1] The list of the physical and perceived factors corresponds to Redman et al. (2013, p. 121).

to a distortion that inhibits a rapid change in mobility behaviour.

In table 2 in the appendix, the modal split for the ten most populous cities in Germany is stated regarding walking, cycling as well as the use of public transport and private motorised vehicles. The mobility pattern is expressed in trips per person.

2.2 Congestion in urban areas

The road network is a scarce resource (Newbery, 1990). Because of limited public budgets, increasing vehicles per inhabitant, long-term capital expenditure, increasing transport performance, public acceptance, a lack of available space and extensive construction periods infrastructure is a scant public good. Congestion is therefore the consequence as an increasing demand meets a fixed supply. The development of congestion for an infrastructural bottleneck is shown in the following figure.

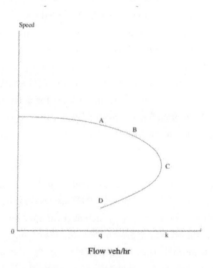

Figure 1: Speed flow relationship for an infrastructural bottleneck
Source: Walters (1961, p. 694)

As the flow increases, the average speed falls at point A below the free flow speed. The bottleneck is completely exhausted at point C with the maximum flow k. This point can be interpreted as a Nash equilibrium since the flow is highest at a certain speed. If the traffic increases to point D, the flow decreases back to q which imposes a lower average speed on the road users. At this point, it would be best for all travellers is the additional road users either change the route or the mode of transport. This situation

3

is considered as congestion. The individual's ambition to choose the fastest alternative, which in this case is the car, leads to a reduction in the average speed of other road users. External costs are thus imposed on other drivers which can be expressed by $\frac{dC}{dq} = c + q\frac{dc}{dq}$ where the marginal costs of the user $(\frac{dC}{dq})$ are the sum of private costs (c) for example fuel or the depreciation of the vehicles, the time dependent costs (C_i) and the marginal costs of the other users for example because of speed reduction (Lindsey & Verhoef, 2000; Newbery, 1990). Direct external costs for other road users arise from opportunity costs due to increased travel times and schedule delay costs due to possible lags (Button, 1990). The marginal costs of the other users can be understood as the social costs of road use which are the sum of accident externalities, environmental pollution, road damage and congestion (Button, 1990; Newbery, 1990).

Jones-Lee & Jones-Lee (1990) examine the accident externalities between 10% and 20% of the total road costs.[2] If the traffic is twice as heavy, the risk of an accident happening to each car is increased by 19%.[3] The environmental pollution contributes 10% to the total road costs (Newbery, 1987). Faiz et al. (1990) examined that at lower speeds vehicles have a higher marginal environmental pollution per additional unit of distance in contrast to higher speeds.

In order to convey the burden of congestion on households and economies in the most understandable way, Cookson (2018) calculated the total economic costs of congestion for the United States, United Kingdom and Germany which were calculated on the basis of of the values for the year 2017 in local currency. Source data were adjusted with a country-specific inflation index if necessary. The direct costs are borne by the driver by using the roads in traffic jams, which include the value or opportunity costs of the time, the user wastes in traffic jams. In addition there are the costs for the additional fuel consumption and the overall social costs for the emissions of his vehicle. The indirect costs are borne by households in the form of higher prices for goods and services that companies demand due to congestion. Cookson (2018) estimates the overall costs of congestion for the most-congested cities in Germany at € 16.40 bn. with an average cost per drive of € 1,770 per year. In table 3 the congestion costs for the ten largest cities in Germany. In absolute terms, external costs are highest in Berlin at € 6.90 bn. due to the large population. The most congested city, on the other hand, is Munich, where the inhabitants spend 16% of their travel time in traffic jams, which results in 51 congestion hours at peak times and € 2,984 per citizen in externalities.

---

[2] Shefer (1994) assumes an inverted u-shaped relationship between road fatalities and the traffic density in urban areas. He argues that the fatalities increase up to a certain point from which of the average driving speed decreases so that the fatalities decrease as well.

[3] Newbery (1990) argue that the accident externalities are covered by insurance companies of the individual causing the accident. Nevertheless, the insurance cover pool has been provided by the public.

2.3 Counter-mechanisms to urban congestion

2.3.1 Demand management in urban transportation systems

The internalisation of externalities caused by congestion is much discussed. Because a constantly increasing demand meets a comparatively inelastic supply, the management of demand is an important aspect in combating congestion in urban areas. This not only involves reducing demand through a price mechanism, but also through measures such as qualitative or quantitative scarcity of supply and campaigns to increase public awareness. The first option is the limitation of infrastructure access through *concessions* (Guasch, 2004), prohibitions (for example with car-free days (Holzapfel, 2016)). Besides the command control there is the possibility of incentivation by moral suasion (Grazi & van den Bergh, 2008; Steg, 2003) or nudging (Lehner et al., 2016). *Moral suasion* can, for example, be generated by public campaigns that raise awareness of congestion problems in order to mobilise the urban population to switch to another mode of transport. Both, moral suasion and nudging, are considered to be short-term effective but lack a long-term positive impact (Marchiori et al., 2017). Since the costs of alternative modes of transport are taken into account by the traveller, the cost reduction of substitutions, for example the public transport, can be reduced in order to achieve a modal shift relieving the congested roads. Besides the provision of stronger subsidised or free public transport (De Witte et al., 2008) or cycling promotion schemes (Uttley & Lovelace, 2016). Redman et al. (2013) emphasise that a modal shift from private car to public transport can be achieved through increasing attractiveness especially in terms of the reliability of public transport. Coase (1960) provides with *negotiations* a solution for the social optimum problem known as the *Coase Theorem*. If one considers a situation of congestion and two individuals, a manager with high costs of time and a worker with low costs of time ($\alpha_{Manager} > \alpha_{Worker}$) it would be rational that the worker reduces his demand for road transport to a point where the manager can travel at free flow speed. Since the manager has a higher value of time, a social optimum is achieved through this negotiation.[4]

Since individuals tend to make more efficient choices when confronted with the social costs and benefits of their choice, it is intuitive to assume that an internalisation of the externalities through a price mechanism. Therefore, congestion charges are introduced in the following chapter.

---

[4] Kahneman et al. (1990) question the practical applicability of the *Coase Theorem* in bilateral bargaining situation because of the *Endowment Effect*. The effect displays the asymmetry in the perception of a fair price: The seller systematically overestimates the price of a product and the buyer systematically underestimates the price. In the negotiation between the worker and the manager, the social optimum does not necessarily has to be achieved because of the bargaining parties does not accept the conditions.

2.3.2 Static and dynamic congestion charges

The price theory of congestion charges is developed by Vickrey (1963). The author emphasises that congestion pricing reflecting the severity of congestion. The charges can be separated in static models (time independent) and dynamic models (time independent) as well as first best pricing and second best pricing (Lindsey & Verhoef, 2000). The basis for the theory is the static model as a first best solution for the problem. Walters (1961) developed an equilibrium model indicating the congestion charge. (Lindsey & Verhoef, 2000) continued to develop the equilibrium model of the generalised costs per trip and the traffic flow by adding the optimum tax level.

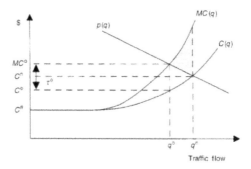

Figure 2: Optimal road pricing in the static model
Source: Lindsey & Verhoef (2001, p. 79)

Lindsey & Verhoef (2001) assume a situation with no infrastructural bottleneck and a free flow (flat function) at the cost $C(q) = C^{ff}$. Comparable with figure 1, the traffic flow increases. Additional costs arise as congestion occurs ($C(q) > C^{ff}$). The flow is interpreted as the quantity of demanded trips. The authors assume a demand curve ($p(q)$) with a downward slope reflecting the individual's preference to make less trips at an increasing price. The *no-toll equilibrium* is the intersection of $C(q)$ and $p(q)$ so that the traffic flow q^n has the costs C^n. If the social optimum is at the Nash equilibrium of q^o trips at the marginal costs MC_q having the optimal costs C^o an optimum tax (congestion charge) would be $\tau^o = MC^o - C^n$. In the first best case the marginal costs in the equilibrium are added to the individual costs of a trip so that the individual pay for the marginal costs of all other individuals. Therefore, the individual cost function from chapter 2.1 is $C_i(t) = \alpha_i T(t) + D_i(t - t^i) + \tau(t)$. In this case, the costs due to the difference between the no-toll traffic flow (q^n) and the traffic flow at the social optimum (q^o) are internalised by the individual traveller. This tax covering the divergence between the individual costs and the social costs is considered as a *Pigou Tax* (Dahlman, 1979). Efficiency gains can

6

be expressed as the increase in social surplus defined by the reduction in total costs minus the reduction in total benefits due to the decrease in traffic. These gains are represented by the shaded area in figure 2. On the other hand, the introduction of the congestion charge result for the users continuing to travel (q_o) with a cost increase of $MC^o - C^n$. The individuals who stop travelling $(q^n - q^o$ have to pay the charge. These losses because of a fixed congestion cost are the origin of opposition to congestion pricing (Lindsey & Verhoef, 2001). Therefore, the dynamic model seems to be more efficient than the static model since the demand for completing trips changes during daytime. Therefore, inefficiencies among the day are considered. The example of commuter traffic shows that the intensity varies during the day. While the load in the morning and afternoon is comparatively high, traffic at other times of the day is moderate. This phenomenon is expressed in the following figure.

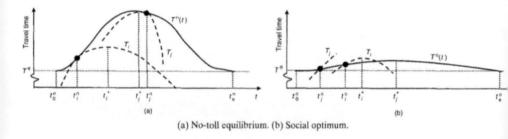

(a) No-toll equilibrium. (b) Social optimum.

Figure 3: Intraday variation of the travel time under the no-toll and social equilibrium
Source: Lindsey & Verhoef (2001, p. 82)

The time dependent model specify how speed and flow evolve over times. Figure 3.(a) emphasises the problem under one daily peak at t^j where due to congestion the travel time is at its maximum. If then the prices are extended considering the intraday peaks, the overall variance in travel time is reduced. An example is 3.(b). The reduced demand is achieved through a modal shift to faster options or to a intraday shift where for example a shopping trip is not conducted during the daily peak but at a less congested time (Xiao et al., 2011, 2012).

The social optimum in figure 3.(b) is considered as an optimal control problem due to a trade-off between schedule delay costs and travel time costs (Lindsey & Verhoef, 2001). The optimal time-varying toll having the proposition to internalise the externalities associated with arrival time is described as a *fine toll*. This toll depends on the joint frequency distribution regarding schedule delay costs, values of time and the preferred arrival times. Hence, the two individuals described in chapter 2.3.1 have an optimum travel time: The manager with a high value of time can travel at the reduced peak t^j in figure 3.(b). He pays the congestion charge as a premium for the shorter travel time and

7

reduced schedule delay costs. In times with no congestion, the individuals do not pay the toll. Hence, the resource are allocated more efficiency because of the reduction in travel time costs due to reordering the the individual's travel time preferences.

Besides the advantages because of congestion charges several constraints arise which are explained in the following chapter.

2.3.3 Constraints in congestion charges

The first constraint of implementing congestion charges are the substantial investments in the systems with operating costs lowering the revenue of the charge by the cost of collection. Furthermore, the speed in the network is reduced because of travelling delays due to stops for payment.

The second constraint are the toll-free alternatives in the network (Lindsey & Verhoef, 2001). The "parallel-routes network problem" is solved by cordon tolling covering large parts of the city. Nonetheless, the congestion charge can cause an overload problem regarding the other modes of transport especially the public transport [5]

The third constraint is the heterogeneity of users since the vehicles differ in terms of occupied road space, environmental pollution (for example because of noise or exhaust gases), the number of people carried as well as the efficiency. Analogous to the *Tinbergen Rule*, the congestion charge aims at the reduction of congestion not on the improvement of the urban environment (Arrow, 1958).

The fourth constraint is the public acceptability and acceptance of road pricing schemes (Jones, 2003).[6] Zheng et al. (2014) emphasise that public support is critical when it comes to the implementation of a congestion charge. Hence, the public acceptance has a high impact on the congestion charge policy. According to Zheng et al. (2014), the introduction is most effectively when the congestion charge is not framed as being imposed because of congestion but due to environmental reasons. The authors admit that a sample drawn in the two cities of Brisbane and Melbourne lacks in terms of external validity since traffic problems are not as severe as in other cities. Regional differences regarding the importance of freedom, the fairness of the charge, the trust in the action of the local authorities / government, awareness of the problems, the perceived effectiveness (ex ante and ex post) as well as the complexity of the charging scheme and the socio-demographic background of the respondent (Kim et al., 2013; Schmöcker et al., 2012). Due to the complexity of public

---

[5] The *Mohring Effect* for public infrastructure leads to an overall improvement of public transport. The *Mohring Effect* is the observation that, if the frequency of a transit service (e.g., buses per hour) increases with demand, then a rise in demand shortens the waiting times of passengers at stops and stations. Because waiting time forms part of the costs of transportation, the The *Mohring Effect* implies increasing returns to scale for scheduled urban transport services (Mohring, 1972).

[6] A detailed record of failed attempts introducing a congestion charge is stated by Hensher & Li (2013).

opinion formation, the attempt to introduce a congestion charge is a risky proposition for political decision-makers, who not only suffer the loss of reputation among their own voters in a failed introduction of the road pricing scheme, but also continue to fall in favour with other voters. A further obstacle to the introduction is the conflict with particular interests. An example of the attempted introduction is New York City. Although 67% of the citizens supported a road charge zone in 2007, it was blocked by the State Legislature because of strong opposition by the motorists (Schaller, 2010). This is an explanation that even if the introduction is associated with social benefits, the congestion charge is rather an exception than a rule.

3. CASE STUDIES

3.1 Introduction to the case studies

In order to derive the recommendations for action for Germany, case studies are used to identify success factors. Examples from London, Stockholm and Singapore are used. Since the cases are enriched with secondary information, the focus in the selection of cities is on a good level of information. Each use case is initiated with the historical development. The pricing scheme is then categorised and analysed for exceptions. The last step is to examine the efficiency of the congestion charge. Efficiency is the assessment criterion that can be used to describe whether a measure is suitable for achieving a given goal in this case, the acceleration of urban traffic to reduce the negative effects on road users as a whole in a certain way. In table 4 in the appendix, the congestion charges are compared.

3.2 London

At the end of the 20th the average travel speed in London was lower than at the end of the 19th century before the car was introduced (Newbery, 1990). In 1964, an urban congestion charge was considered first by the Ministry of Transport (Smeed, 1964). At the average workday, one million travellers entered central London between 07:00 and 10:00 where 30% used their car (Leape, 2006). The basis for the introduction of the congestion charge in London was excessive public consultation in the period of 18 months (Bhatt et al., 2008; Santos, 2008). In 1999, 90% of the surveyed residents thought that there is too much traffic in London. 41% of the respondents thought that the best way of funding public transport is through a congestion charge. (Santos, 2008).

The congestion charge was introduced in February, 2003 with a daily charge of £5.00. Preliminary simulations have shown that the effectiveness of the congestion charge is based on two factors: Firstly, the location of the cordons, i.e. the zone in which the charge is levied. On the other hand, the level of costs incurred in collecting the fee where the pricing scheme is a crucial factor. Variable charges are considered to have a reduced efficency because of their complicated nature (Santos et al., 2001). Within the charging area, the fee is collected between 07:00 a.m. and 06:30 p.m. from Monday to Friday. The charge is imposed on vehicles driving or parking within central London. Residents living withing the zone receive a discount of 90%. Furthermore, registered cars which emit carbon dioxide of $< 75g/km$ meeting the Euro 5 emission standard or are either electric or plug-in hybrid electric receive a 100% discount. Between 30% and 40% are exempt because of the 100% reduction. 10% of the vehicles receive resident's discounts (Evans, 2008). The congestion charge was increased on July, 2005 to £8.00, on January, 2011 to £10.00 and finally in June, 2014 to £11.50. The enforcement of the congestion charge is secured via video cameras and automatic number plate registration. If the vehicle is not registered, the user has to pay £65 within 14 days, after these days £195.50 and after 28 days legal issues will be initiated. In 2008, the *Low Emission Zone (LEZ)* was introduced where vehicles with certain emission standards have to pay £12.50. Furthermore, trucks and busses have to pay up to 100£ when entering the *LEZ*. In the *Ultra Low Emission Zone (ULZ)* all vehicles are urged to pay the charges from the Low Emission Zone. In contrast to the regular toll, the *LEZ* and *ULZ* are collected anytime (TfL, 2019).

The short-term impact of the congestion charge in the charging zone between 2002 (before) and 2003 (after the introduction) is reported by Leape (2006): The overall transport performance was reduced by 12% while the transport performance of cars decreased by 34%. The performance of buses increased by 21% and for bicycles by 28%. Leape (2006) reports that the average speed per hour increased from 14.1 km/hour (no-toll), to 16.4 km/hour (£5.00) up to 16.9 km/hour (£8.00). The introduction of the congestion charge had a significant negative impact on the urban retail business (Leape, 2006). A survey among the 500 firms in London reports that 72% of the business felt that the charging scheme was working while 14% said that it was a failure. 58% of the companies thought that the congestion charge improved London's image (Clark, 2004). Quddus et al. (2007a) estimate a reduction of sales between 5.5% and 8.2% in central London which depends on the accessibility of the locations: In areas which were accessed mostly by cars the reduction of sales is estimated at 10% whereas between 3% and 6% of the shopping trips were absolved with the private car (Quddus et al., 2007b).

The social benefits minus the social costs is estimated at £67 mio. for the year 2005. A detailed calculation taken from Leape (2006) is displayed in table 1 in the appendix. The measurement of the impact is difficult since the effects of the congestion charge cannot be separated from the other effects. The ceteris paribus assumption can for example be offset by an increased efficiency of the vehicles.

3.3 Stockholm

Road pricing schemes have triggered an intense debate in Sweden since the 1990s. The election victories of the Greens have led to an intensification of the debate and initial feasibility studies (Zheng et al., 2014). The congestion charge in Stockholm was introduced in January 2006 and closed down in July 2006 at the end of a trial period. This period was accompanied by a public information campaign and public consulting where the inhabitants had the opportunity to participate in shaping the fee according to their type of levy. The amount was explained by the company surveyed, in which social benefits were explained in detail. Before and during the test phase, the residents had the opportunity to obtain advice on the Congestion Charge from information centres (Schuitema et al., 2010). At this time, test phases for congestion charges were initiated not only in Stockholm, but throughout Sweden. The test period ended with a national referendum in which residents could decide on the introduction of congestion pricing in their respective cities. Stockholm was the only city in which the referendum was successful, which is why a peak-based congestion charge was introduced in August 2007 (Eliasson, 2009). The introduction of the congestion charge was accompanied by an extension of the public transport supply. In contrast to the flat-charge in London, the price for congestion is peak based. Figure 5 in the appendix displays the intraday variation of the congestion charge. The fee is due each time you enter or leave the zone and is limited to € 5.57 (SEK60) per vehicle per day. 18 control points are located at Stockholm city entrances and exits where 60% of the charge is collected automatically through number plate registration through cameras. 40% of the payments are made outside the zone at local shops or through bank transfers (Eliasson, 2009). The congestion charge is only levied between Monday and Friday whereas no collection happens during the weekends and national holidays (Eliasson, 2009).

The wide public communication of the reasons of the introduction of the charge resulted in higher acceptance after the introduction than during the referendum (Schuitema et al., 2010). In contrast to the explanation in chapter 2.3.3, Eliasson & Jonsson (2011) argues that in the example of Stockholm overcoming uncertainty about personal effects was not the main reason for the increased acceptance of the charge. The first important factor was the result of the introduction: The advantages regarding the increased efficiency were constantly communicated to the population (Eliasson & Jonsson, 2011). Decisive here, however, were less the objective than the subjective and perceived advantages (Winslott-Hiselius et al., 2009). According to Schuitema et al. (2010), the second important factor is the high sensitivity for environmental issues of the Swedish population. Besides the Swedes in general are willing to give up their car (Eliasson & Jonsson, 2011). The fact that the cost of using their own car increased on the one hand and the quality of public transport

11

improved on the other facilitated the change: This is reflected in the fact that the modal split of public transport increased by 4 percentage points. In addition, the number of vehicles entering the city centre has decreased by 22% compared to 2005 (Schuitema et al., 2010). Börjesson et al. (2012) were able to show that the reduction in the use of one's own car was by no means a short-term effect, but rather that the reduction of 20% in entering the cordon remained relatively constant between 2006 and 2011.[7] It can therefore be assumed that the urban population has become accustomed to the congestion charge. Moreover, the CO^2 emissions were reduced by 14% in the city centre. Outside of the cordon, CO^2 emissions decreased by 2.5% (Eliasson, 2014).

3.4   Singapore

Singapore is characterised by a high degree of urbanisation and an economic upswing since the 1960s. To increase efficiency, a license-based system to reduce urban traffic was introduced in 1975. The *Singapore Area Licensing Scheme (ALS)* is considered as the first urban congestion pricing (Phang & Toh, 1997). The ALS covered the central business district and was based on a daily fee which increased constantly over the years (Gomez-Ibanez et al., 1994). Phang & Toh (1997) postulates that the traffic during the morning peak hours was reduced by 45% and the average speed increased from 19km/h before the ALS to 36km/h after the introduction. The ALS which was operated between 1975 and 1998 was comparable with the congestion charge in London.

In 1998, the Electronic Road Pricing (ERP) was introduced detaching the ALS. The ERP is a gantry-based system where vehicles have to be equipped with an on-board-unit which interacts with an exterior system of radio beacons mounted on gantries (Santos, 2008). 93 gentries are distributed over the city separating the central district into different cordons. The number of vehicles entering the cordons is tracked by the gentries and the corridor charge is adjusted every quarter depending on the congestion within the cordons (Olszewski & Xie, 2005). In August 2019, the corridor charge is between € 0.30 (SGD0.50) and € 2.00 (SGD3.00) per entry (Liew, 2019). The intraday variation between the gentries is adjusted that citizens are able to travel between 20 and 30km/h in the congestion zone and between 45 and 65 km/h on motorways (Metz, 2018). The ERP achieved a reduction in traffic volumes in the central business district between 10 and 15% (Chin, 2009).

In 2020, the distance-based charging system *ERP2* will be introduced covering the entire area of Singapore. A Global Navigation Satellite System will replace the on-board units and a possible removal of the gentries is discussed having the potential of decreasing the cost of collection while increasing the efficiency of the system (MTR, 2019). Furthermore,

---

[7]Börjesson & Kristoffersson (2018) extended the study and examined the same constant effect from 2006 to 2016.

the new on-board unit has the purpose of a platform integrating traffic information and parking management (TheStraitsTimes, 2019).

Besides the road pricing system, the car ownership is limited by a licence bidding system which costs about €52,000 (SGD80,000) for ten years (Metz, 2018). The high operating costs because of the road pricing together with the licence decreased car usage and promoted public transport.

4. RECOMMENDATIONS FOR GERMANY

SpiegelOnline reports in a survey that the majority of Germans is against a city toll. 56.7% of the respondents had a negative or rather negative view on congestion charge. Furthermore, the acceptance of the urban pricing scheme correlated negatively with the age and positively with population density of the city in which the respondents lived (SpiegelOnline, 2019). Nonetheless, five success factors can be derived from the case studies.

1. Efficient congestion charging system: The experiences from Singapore between 1975 and 1998 have shown, that a paper based collection or a toll booth systems is associated with waiting times when entering the charging zone which reduces the intended impact. The example of Singapore shows that a system close to a first-best solution is possible considering a distance-based charging system within the cordons. The operating costs are reduced because of the replacement of physical gantries by satellite navigation policy. The charge is adjusted to the distance as well as the amount of vehicles in the cordon whereas an efficient scheme can be operated. Effective operations of the systems are guaranteed if the agencies collecting the charge manage the whole process from the registration of the vehicles to the collection of the charge. The *LEZ* and especially the *ULEZ* provide an example of addressing the heterogeneity of the vehicles. If the congestion charge is considered as an environmental charge, higher pollution levels are addressed with a surcharge. The first recommendation is the introduction of an *efficient distance-based congestion charging system with a minimum of hardware addressing the heterogeneity of the vehicles.*

2. Public acceptance: As the congestion charge was introduced successfully, Stockholm was the only major city in Sweden in which the referendum was in favour of the charge. Hysing (2015) describes this problem as a dilemma between the common good (reduction of externalities) and the individual good (no internalisation of externalities). The simplicity of communication, both in terms of reasons and objectives, is also crucial when it comes to raising public awareness During the trial period in Stockholm the citizens had the opportunity to get used to the congestion charge. In Singapore, the whole costs of the public good *infrastructure* are paid by the driver, in the other examples only the marginal costs. The democratic examples of London and Stockholm have shown the importance of public consulting and information. In both cases, the introduction was accompanied by a long planning process guaranteeing transparency for the stakeholders for the whole time. The public communication has to ensure that the perceived advantages are in line with the objective advantages. In chapter 2.3.3 the congestion charge is considered as being more effective when it is framed not only against congestion but for the environment. Hysing (2015) adds that the introduction of congestion charges is most successful when being included into an infrastructure package. Therefore, the second recommendation is that the charge has to be *promoted excessively, timely and transparently as an environmental charge being part of an infrastructure package with a trial period and the possibility of overcoming uncertainty.*

3. Promotion of modes of transport: The introduction of a congestion charge initiated a modal shift resulting in an increasing demand of different modes of transport. In the three examples from chapter 3, the revenues of the system were used for public transport subsidies. Social sustainability is further addressed by subsidies for public transport. The example concerning the value of time in chapter 2.1 the user of public transport are rewarded with reduced prices or increased quality of public transport. Because of considered network effects a strong modal shift from private motorised car to public transport results in a congestion shift. Therefore, the third recommendation is the *promotion of alternative modes of transport and using the congestion charge for investments in the public infrastructure.*

4. Social sustainability and equity effects: Even though the reduced externalities exceed the costs of the congestion charge, the introduction is considered to cause winners and losers since the reduction of externalities does not necessarily has an impact on distributional aspects (Richardson et al., 2010). Therefore, the social compatibility of a city toll is an important core element. Singapore where the congestion charge is used for reducing the tax of low-emission vehicles, is a good example of the re-distributional mechanism of a congestion charge.The use case of London shows that various exceptions of the charge contribute to a social sustainable system. The fourth recommendation is *maintenance of social sustainability by redistributing the benefits of the congestion charge to the citizens.*

5. Political leadership: Congestion charges embedded in an integrated transport policy need a strong political leadership. The example of New York City from chapter 2.3.3 emphasises that public consultation does not necessarily protects against a failure of introducing the congestion charge. Arnold et al. (2010) argue that only because of the strong political leadership in London and Stockholm, the implementation was successful. The executive mandate for the congestion charge has to be supported by a vision, transparent information and open discussions. In this case, democratic decision-making is based on a continuous continuation and a cross-party will to introduce it. In the rarest of cases, the process from the introduction of the congestion charge from a feasibility study through conception and public information campaigns to a successful introduction within a legislative period can be accomplished. The cross-party will can be generated by a strong public opinion, which arises from sensitivity on the part of the public as well as from an attitude of the population that is congruent with the goals of congestion charge. Therefore, the fifth recommendation is *establishing a cross-party commitment where the interests of all stakeholders are considered.*

5. REFLECTION AND CONCLUSION

The collection of a urban congestion charge would be a real innovation in German transport policy. A possible concrete form could be a distance-based system accompanied by extensive public as well as increased investments in public transport. Experience from Stockholm and London suggest that after overcoming initial resistances, an urban charging scheme can be operated comparatively 'inconspicuously' and cause permanent changes in the behaviour of road users. If the political objective is to significantly reduce vehicle traffic in the city centre - primarily for ecological reasons - there are only limited alternatives to congestion charges. These mainly consist of regulatory traffic restrictions which, although they have lower administration costs than a city toll, cannot generate any additional funds to improve local public transport and also disregard the different willingness to pay for the use of the public good. A congestion charge would make the cost of a car trip into the city more expensive compared to the alternatives. This would mean that people who are not necessarily dependent on a car would increasingly switch to other means of transport or use the car more effectively, for example by forming car pools. Nevertheless, nobody who is dependent on the car would be banned from driving in the cities. The question of an increased acceptance by the public of a charge with distributional effects needs a further academic discussion.

The introduction of a city toll would be an economically and ecologically sensible response to the many problems associated with increasing car traffic in German cities. With a trial - initially in individual model areas - the politicians would set out on the path of a modern transport policy. However, it does not stop with the congestion charge. Rather, alternatives such as better local public transport, the expansion of ride-sharing and car-sharing services, and incentive systems for cyclists and pedestrians would have to be promoted.

Appendix

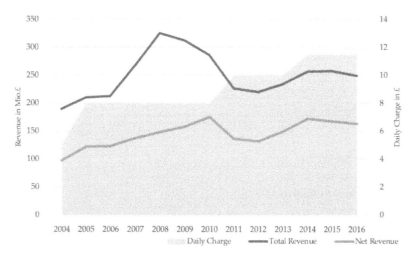

Figure 4: Development of the London congestion charge between 2004 and 2016
Source: Own presentation, the revenue was taken from the annual accounts of TfL

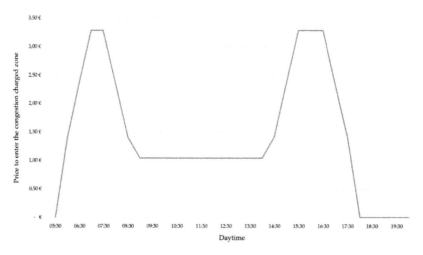

Figure 5: Intraday variation of the Stockholm congestion charge
Source: Own representation; exchange rate: SEK to EUR: 0.094

Annual costs	2005
Individual costs of the charge	210
Transport for London administrative costs	5
Scheme operation costs	85
Setup costs (opportunity costs + depreciation)[8]	23
Traffic management costs[9]	20
Charge-payer compliance costs	30
Total annual costs	**373**

Annual benefits	2005
Revenue from the charge	210
Time savings and reliability benefits:	
Car users	110
Vans, trucks	35
Taxis	40
Buses	42
Deterred drivers	-25
Reduced accidents	15
Reduced CO_2 emissions	3
Other resource savings	10
Total annual benefits	**440**

Table 1: Estimates of the social benefits and costs of the London congestion charging scheme

Source: (Leape, 2006)

---

[8]Leape (2006) calculates the set up costs between 2000 and 2003 at £170 million with a depreciation of 10% at a discount rate of 3.5%.

[9]Primarily increased bus services

Rank	City	Inhabitants	Area	Population density	Walking	Cycling	Public transport	PMV	Year
1	Berlin[10]	3,644,826	801.68	4,088	28%	13%	27%	32%	2014
2	Hamburg[11]	1,841,179	755.22	2,438	28%	12%	18%	42%	2014
3	Munich[12]	1,481,508	310.70	4,736	27%	17%	23%	33%	2008
4	Cologne[13]	1,085,664	405.02	2,681	28%	15%	15%	42%	2008
5	Frankfurt[14]	753,056	248.31	3,033	33%	12%	22%	33%	2013
6	Stuttgart[15]	634,830	207.35	3,062	26%	6%	26%	42%	2010
7	Duesseldorf[16]	619,294	217.41	2,849	31%	13%	22%	34%	2014
8	Leipzig[17]	587,857	297.80	1,974	25%	17%	18%	40%	2015
9	Dortmund[18]	587,010	280.71	2,091	27%	6%	20%	47%	2013
10	Essen[19]	583,109	210.34	2,772	22%	5%	19%	54%	2011
	Weighted average				28%	13%	22%	38%	

Table 2: Modal split in selected German cities

Source: Own representation

[10]https://www.berlin.de/senuvk/verkehr/politik_planung/zahlen_fakten/download/SrV_2013_Berlin_Steckbrief.pdf
[11]https://www.bundestag.de/resource/blob/535044/f9877fd834da2c1bf7c7bb02299da09e/%20wd-5-084-17-pdf-data.pdf
[12]https://www.bundestag.de/resource/blob/535044/f9877fd834da2c1bf7c7bb02299da09e/%20wd-5-084-17-pdf-data.pdf
[13]https://www.bundestag.de/resource/blob/535044/f9877fd834da2c1bf7c7bb02299da09e/%20wd-5-084-17-pdf-data.pdf
[14]https://www.frankfurt.de/sixcms/media.php/738/Frankfurt%20am%202013%20Main%202013%20Steckbrief%20alle%20Tage.pdf
[15]http://www.vvs.de/download/Mobilitaetsbroschuere.pdf
[16]https://www.bundestag.de/resource/blob/535044/f9877fd834da2c1bf7c7bb02299da09e/%20wd-5-084-17-pdf-data.pdf
[17]https://static.leipzig.de/fileadmin/mediendatenbank/leipzig-de/Stadt/02.6_Dez6_Stadtentwicklung_Bau/66_Verkehrs_und_Tiefbauamt/SrV-2015-Information-zu-Kennziffern-der-Mobilitat-fur-die-Stadt-Leipzig.pdf
[18]https://www.dortmund.de/media/p/stadtplanungs_und_bauordnungsamt/stadtplanung_bauordnung_downloads/verkehrsplanung/Vorlage_Mobilitaetsverhalten.pdf
[19]https://media.essen.de/media/wwwessende/aemter/61/dokumente_7/verkehrsthemen/Masterplan_Verkehr_Essen_2018.pdf

Rank	City	Inhabitants	Area	Population density	Congestion rate[20]	Costs/driver (EUR)	Costs (MEUR)
1	Berlin	3,644,826	801.68	4,088	14%	2,811	6,900
2	Hamburg	1,841,179	755.22	2,438	14%	2,646	3,500
3	Munich	1,481,508	310.70	4,736	16%	2,984	2,900
4	Cologne	1,085,664	405.02	2,681	11%	2,107	1,400
5	Frankfurt	753,056	248.31	3,033	10%	1,820	910
6	Stuttgart	634,830	207.35	3,062	13%	2,386	920
7	Duesseldorf	619,294	217.41	2,849	10%	1,823	770
8	Leipzig	587,857	297.80	1,974	n.a.	n.a.	n.a.
9	Dortmund	587,010	280.71	2,091	10%	2,129	2,200
10	Essen	583,109	210.34	2,772	10%	2,129	2,200

Table 3: Congestion Externalities

Source: (Cookson, 2018)

[20]Share of travel time which is spent in traffic jams.

	London	Stockholm	Singapore
Purpose	Congestion management, transit promotion, emission reduction	Congestion management, transit promotion, emission reduction	Congestion management, transit promotion
Pricing scheme	Urban area pricing	Urban cordon pricing, peak pricing with various reductions	Cordon with time-of-day pricing in city centre & urban motorways, intraday variation & quarterly review
Technology	Automated number plate registration	Automated number plate registration	Short-range communication with gentries, onboard units & automated number plate registration for enforcement
Impact	25% reduction in traffic, modal shift to public transport & cycling, problems for urban business examined	20% reduction in traffic, 10-14% decrease in emissions, 2-10% improvement in air quality	Achievement of target speeds in the desired band
Initial investment	€190 mio.	€186 mio.	€132 mio.
Annual operating costs	€151 mio.	€9 mio.	€17 mio.
Annual net revenue	€160 mio.	€122 mio.	€100 mio.
Investments	80% of the revenue is used for transit & 20% for transportation	Funding transportation and transit improvements	Revenues are returned to vehicle owners through tax rebates

Table 4: Road pricing schemes in London, Stockholm and Singapore

Source: Own representation, data derived from (Arnold et al., 2010; Provonsha, 2017)

REFERENCES

Arnold, R., Smith, V. C., Doan, J. Q., Barry, R. N., Blakesley, J. L., DeCorla-Souza, P. T., Muriello, M. F., Murthy, G. N., Rubstello, P. K., Thompson, N. A., et al. (2010). *Reducing congestion and funding transportation using road pricing in Europe and Singapore*. Technical report, United States. Federal Highway Administration.

Arrow, K. J. (1958). Tinbergen on economic policy. *Journal of the American Statistical Association*, 53(281), 89–97.

Bhatt, K., Higgins, T., Berg, J. T., Analytics, K., et al. (2008). *Lessons learned from international experience in congestion pricing*. Technical report, United States. Federal Highway Administration.

Börjesson, M., Eliasson, J., Hugosson, M. B., & Brundell-Freij, K. (2012). The stockholm congestion charges—5 years on. effects, acceptability and lessons learnt. *Transport Policy*, 20, 1–12.

Börjesson, M. & Kristoffersson, I. (2018). The swedish congestion charges: Ten years on. *Transportation Research Part A: Policy and Practice*, 107, 35–51.

Button, K. (1990). Environmental externalities and transport policy. *Oxford Review of Economic Policy*, 6(2), 61–75.

Chin, K.-K. (2009). *The Singapore experience: The evolution of technologies, costs and benefits, and lessons learnt*. Technical report, OECD/ITF Joint Transport Research Centre Discussion Paper.

Clark, A. (2004). London companies learn to love congestion charge. *The Guardian*, 16.

Clifton, K. J. (2004). Mobility strategies and food shopping for low-income families: A case study. *Journal of Planning Education and Research*, 23(4), 402–413.

Coase, R. H. (1960). The problem of social cost. *The Journal of Law and Economics*, 3, 87–137.

Cookson, G. (2018). Inrix global traffic scorecard. *INRIX Research*.

Dahlman, C. J. (1979). The problem of externality. *The Journal of Law and Economics*, 22(1), 141–162.

De Witte, A., Macharis, C., & Mairesse, O. (2008). How persuasive is 'free'public transport?: a survey among commuters in the brussels capital region. *Transport Policy*, 15(4), 216–224.

Donald, I. J., Cooper, S. R., & Conchie, S. M. (2014). An extended theory of planned behaviour model of the psychological factors affecting commuters' transport mode use. *Journal of Environmental Psychology*, 40, 39–48.

Eliasson, J. (2009). A cost–benefit analysis of the stockholm congestion charging system. *Transportation Research Part A: Policy and Practice*, 43(4), 468–480.

Eliasson, J. (2014). The stockholm congestion charges: an overview. *Stockholm: Centre for Transport Studies CTS Working Paper*, 7, 42.

Eliasson, J. & Jonsson, L. (2011). The unexpected "yes": Explanatory factors behind the positive attitudes to congestion charges in stockholm. *Transport Policy*, 18(4), 636–647.

Evans, R. (2008). *Demand elasticities for car trips to central London as revealed by the Central London Congestion Charge*. Technical report.

Faiz, A., Sinha, K., Walsh, M., & Varma, A. (1990). *Automotive air pollution : issues and options for developing countries*. WPS 492. Washington, DC: The World Bank, Infrastructure and Urban Development Department: August 1990.

Flyvbjerg, B., Skamris Holm, M. K., & Buhl, S. L. (2004). What causes cost overrun in transport infrastructure projects? *Transport Reviews*, 24(1), 3–18.

Gomez-Ibanez, J. A., Gomez-Ibanez, J. A., & Small, K. A. (1994). *Road pricing for congestion management: A survey of international practice*, volume 210. Transportation Research Board.

Grazi, F. & van den Bergh, J. C. (2008). Spatial organization, transport, and climate change: Comparing instruments of spatial planning and policy. *Ecological Economics*, 67(4), 630–639.

Guasch, J. L. (2004). *Granting and renegotiating infrastructure concessions: doing it right*. The World Bank.

Hensher, D. A. & Li, Z. (2013). Referendum voting in road pricing reform: A review of the evidence. *Transport Policy*, 25, 186–197.

Hine, J. & Mitchell, F. (2017). *Transport disadvantage and social exclusion: exclusionary mechanisms in transport in urban Scotland*. Routledge.

Holmgren, J. (2007). Meta-analysis of public transport demand. *Transportation Research Part A: Policy and Practice*, 41(10), 1021–1035.

V

Holzapfel, H. (2016). Mobilitätszukunft: Bewusstseinswandel oder Technik? In *Urbanismus und Verkehr* (pp. 109–112). Springer.

Hysing, E. (2015). Citizen participation or representative government–building legitimacy for the gothenburg congestion tax. *Transport Policy*, 39, 1–8.

Jones, P. (2003). Acceptability of road user charging: meeting the challenge. In J. Schade & B. Schlag (Eds.), *Acceptability of Transport Pricing Strategies* (pp. 27–62). Pergamon Press.

Jones-Lee, M. W. & Jones-Lee, M. (1990). The value of transport safety. *Oxford Review of Economic Policy*, 6(2), 39–60.

Kahneman, D., Knetsch, J. L., & Thaler, R. H. (1990). Experimental tests of the endowment effect and the coase theorem. *Journal of Political Economy*, 98(6), 1325–1348.

Kim, J., Schmöcker, J.-D., Fujii, S., & Noland, R. B. (2013). Attitudes towards road pricing and environmental taxation among us and uk students. *Transportation Research Part A: Policy and Practice*, 48, 50–62.

Klinger, T., Kenworthy, J. R., & Lanzendorf, M. (2013). Dimensions of urban mobility cultures–a comparison of german cities. *Journal of Transport Geography*, 31, 18–29.

Klöckner, C. A. & Matthies, E. (2004). How habits interfere with norm-directed behaviour: A normative decision-making model for travel mode choice. *Journal of Environmental Psychology*, 24(3), 319–327.

Leape, J. (2006). The london congestion charge. *Journal of Economic Perspectives*, 20(4), 157–176.

Lehner, M., Mont, O., & Heiskanen, E. (2016). Nudging–a promising tool for sustainable consumption behaviour? *Journal of Cleaner Production*, 134, 166–177.

Liew, E. (2019). New erp rates for august, 2019. *https://blog.moneysmart.sg/transportation/erp-rates-gantry-singapore/*. Accessed: 2019-10-18.

Lindsey, C. R. & Verhoef, E. T. (2000). *Traffic congestion and congestion pricing*. Technical report, Tinbergen Institute Discussion Paper.

Lindsey, R. & Verhoef, E. (2001). Traffic congestion and congestion pricing. In K. J. Button & D. A. Hensher (Eds.), *Handbook of Transport Systems and Traffic Control* (pp. 77 – 105). Emerald Group Publishing Limited, Bingley.

Marchiori, D. R., Adriaanse, M. A., & De Ridder, D. T. (2017). Unresolved questions in nudging research: Putting the psychology back in nudging. *Social and Personality Psychology Compass*, 11(1), 1–13.

Metz, D. (2018). Tackling urban traffic congestion: The experience of london, stockholm and singapore. *Case Studies on Transport Policy*, 6(4), 494–498.

Mohring, H. (1972). Optimization and scale economies in urban bus transportation. *The American Economic Review*, 62(4), 591–604.

MTR (2019). Erp - next generation erp system. *https://www.mot.gov.sg/about-mot/land-transport/motoring/erp*. Accessed: 2019-10-26.

Newbery, D. M. (1987). *Road User Charges and the Taxation of Road Transport*. IMF Working Paper.

Newbery, D. M. (1990). Pricing and congestion: Economic principles relevant to pricing roads. *Oxford Review of Economic Policy*, 6, 22–38.

Olszewski, P. & Xie, L. (2005). Modelling the effects of road pricing on traffic in singapore. *Transportation Research Part A: Policy and Practice*, 39(7-9), 755–772.

Phang, S.-Y. & Toh, R. S. (1997). From manual to electronic road congestion pricing: The singapore experience and experiment. *Transportation Research Part E: Logistics and Transportation Review*, 33(2), 97–106.

Pilegaard, N. & Fosgerau, M. (2008). Cost benefit analysis of a transport improvement in the case of search unemployment. *Journal of Transport Economics and Policy*, 42(1), 23–42.

Provonsha, E. (2017). *Road pricing in London, Stockholm and Singapore - A way forward for New York City*. Technical report, Tri-State Transportation Campaign - Mobilizing the Region.

Quddus, M. A., Bell, M. G., Schmöcker, J.-D., & Fonzone, A. (2007a). The impact of the congestion charge on the retail business in london: An econometric analysis. *Transport Policy*, 14(5), 433–444.

Quddus, M. A., Carmel, A., & Bell, M. G. (2007b). The impact of the congestion charge on retail: the london experience. *Journal of Transport Economics and Policy*, 41(1), 113–133.

Redman, L., Friman, M., Gärling, T., & Hartig, T. (2013). Quality attributes of public transport that attract car users: A research review. *Transport Policy*, 25, 119–127.

Richardson, T., Isaksson, K., & Gullberg, A. (2010). Changing frames of mobility through radical policy interventions? the stockholm congestion tax. *International Planning Studies*, 15(1), 53–67.

Rietveld, P. (1994). Spatial economic impacts of transport infrastructure supply. *Transportation Research Part A: Policy and Practice*, 28(4), 329–341.

Ryff, C. D. (1989). Happiness is everything, or is it? explorations on the meaning of psychological well-being. *Journal of Personality and Social Psychology*, 57(6), 1069.

Santos, G. (2008). The london congestion chargin scheme. In H. W. Richardson & C.-H. C. Bae (Eds.), *Road congestion pricing in Europe: Implications for the United States*. Edward Elgar Publishing.

Santos, G., Newbery, D., & Rojey, L. (2001). Static versus demand-sensitive models and estimation of second–best cordon tolls: An exercise for eight english towns. *Transportation Research Record*, 1747, 44–50.

Schaller, B. (2010). New york city's congestion pricing experience and implications for road pricing acceptance in the united states. *Transport Policy*, 17(4), 266–273.

Schmöcker, J.-D., Pettersson, P., & Fujii, S. (2012). Comparative analysis of proximal and distal determinants for the acceptance of coercive charging policies in the uk and japan. *International Journal of Sustainable Transportation*, 6(3), 156–173.

Schuitema, G., Steg, L., & Forward, S. (2010). Explaining differences in acceptability before and acceptance after the implementation of a congestion charge in stockholm. *Transportation Research Part A: Policy and Practice*, 44(2), 99–109.

Shefer, D. (1994). Congestion, air pollution, and road fatalities in urban areas. *Accident Analysis & Prevention*, 26(4), 501–509.

Smeed, R. (1964). *Road pricing: the economic and technical possibilities*. Technical report.

SpiegelOnline (2019). Mehrheit der Deutschen lehnt City-Maut ab. *https://www.spiegel.de/auto/aktuell/ citymaut-deutsche-lehnen-innenstadtgebuehr-mehrheitlich-ab-a-1264962.html*. Accessed: 2019-10-26.

Stanley, J. K., Hensher, D. A., Stanley, J. R., & Vella-Brodrick, D. (2011). Mobility, social exclusion and well-being: Exploring the links. *Transportation Research Part A: policy and practice*, 45(8), 789–801.

Steg, L. (2003). Can public transport compete with the private car? *Iatss Research*, 27(2), 27–35.

TfL (2019). Low emission zone and ultra low emission zone. *https://tfl.gov.uk/modes/driving/low-emission-zone*. Accessed: 2019-10-26.

TheStraitsTimes (2019). Erp 2.0 goes the distance with new tech. *https://www. straitstimes.com/singapore/transport/erp-20-goes-the-distance-with-new-tech*. Accessed: 2019-10-26.

Uttley, J. & Lovelace, R. (2016). Cycling promotion schemes and long-term behavioural change: A case study from the university of sheffield. *Case Studies on Transport Policy*, 4(2), 133–142.

Vickrey, W. S. (1963). Pricing in urban and suburban transport. *The American Economic Review*, 53(2), 452–465.

Vickrey, W. S. (1969). Congestion theory and transport investment. *The American Economic Review*, 59(2), 251–260.

Walters, A. A. (1961). The theory and measurement of private and social cost of highway congestion. *Econometrica: Journal of the Econometric Society*, (pp. 676–699).

Winslott-Hiselius, L., Brundell-Freij, K., Vagland, Å., & Byström, C. (2009). The development of public attitudes towards the stockholm congestion trial. *Transportation Research Part A: Policy and Practice*, 43(3), 269–282.

Xiao, F., Shen, W., & Zhang, H. M. (2012). The morning commute under flat toll and tactical waiting. *Transportation Research Part B: Methodological*, 46(10), 1346–1359.

Xiao, F. E., Qian, Z. S., & Zhang, H. M. (2011). The morning commute problem with coarse toll and nonidentical commuters. *Networks and Spatial Economics*, 11(2), 343–369.

Zheng, Z., Liu, Z., Liu, C., & Shiwakoti, N. (2014). Understanding public response to a congestion charge: A random-effects ordered logit approach. *Transportation Research Part A: Policy and Practice*, 70, 117–134.

9 783346 441676